DENTISTRY
Demystified

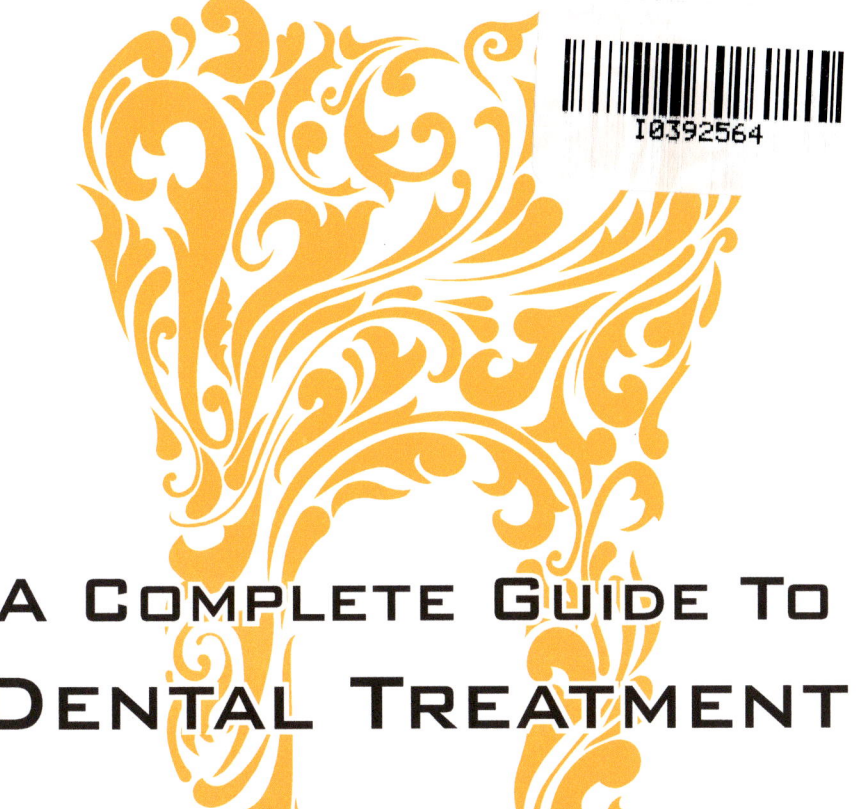

I0392564

A COMPLETE GUIDE TO
DENTAL TREATMENT

DR. NADIM MAJID
BDS, MFGDP (UK) RCS NLP PRACTITIONER

Legal Disclaimer:

The information in this book has been sourced from some of the most credible sources including scientific research and reports. However IT IS NOT INTENDED TO PROVIDE MEDICAL ADVICE. *Neither the author nor the publisher of this book accepts any liability or responsibility for the misuse or abuse of the information provided, especially those pertaining to dental* treatment, procedure, and application of medication.

The author and publisher shall in no event be held liable to any party for any direct, indirect, punitive, special, incidental or other consequential damages arising directly or indirectly from any use of this material, which is provided "as is," and without warranties.

Copyright:

The author of this book **Dentistry Demystified** has been confirmed as **Dr Nadim Majid, Lifestyle Dental Clinic.** No one is allowed to reproduce in either full or part, in any format including print and electronic mediums, audio files and video files, without the express permission from the author and/or publishers.

©2012 Dr. Nadim Majid, Lifestyle Dental Clinic

TABLE OF CONTENTS

Page

What Others Are Saying About the Author

"I am very fortunate to know Nadim who has two powerful characteristics that take him to the front in his profession. This guy is an innovator, bringing new ideas to the dental field with particular strengths in communication and marketing. Add this to him being an all round nice guy and we have a dentist that you would be foolish not to keep an eye on."

Michael Oliver, Principal and Director, Olivers Dental Studio

"Nadim is a competent and dedicated dental professional. He has a flair for marketing but a passion for dentistry. He is focused on acheiving the best possible results for his patients. The quality of his clinical work is of the highest standards. Nadim invests in building a team of carers who will help patients get the very best from their smile."

David Holland, Owner, EXELA Limited

"Nadim is clearly a forward-thinking dentist and entrepreneur who sets the highest standards for himself and his team. I have been impressed by his openness and willingness to share his knowledge and experience with others. He works hard on himself first and foremost, committed to constantly improving his skills both professionally and in his personal life. All in all, Nadim is someone whose opinion matters to his peers and I personally highly value his thoughts, views and his contributions."

Lubna Alam-Orths, Owner, Practice Principal and Dentist, True Dentistry

"Nadim is a technology/systems entrepreneur and trendsetter. This has helped him to setup a business from scratch and constantly improve efficiency and service to his clients. Nadim also looks at the bigger picture – mind, body, soul, for the people around him, which makes him a very approachable and liked personality. He is one of the most open minded and helpful people I know personally and has been the inspiration to many positive influences for me and my business."

Jan Einfeldt, Owner, Staplehurst Dental Practice

"Nadim lives his life to the full. He enjoys seeing people succeed and supporting people and patients through their journey. He is very knowledgeable about the business of dentistry and happy to share that knowledge. He is a pleasure to know and work with."

Mhari Coxon, Motivational/Inspirational/Implementation Speaker, Professional Speaker

INTRODUCTION

If you are looking for more information about dentistry or a specific dental procedure, you have come to the right place. My name is Dr. Nadim Majid and I am the Dental Surgeon and CEO at Lifestyle Dental Clinic located in Preston Fulwood. As a dentist I believe that the most important aspect of the profession is communication. The ability to communicate the dental treatment to the patient in a manner they understand and then offer them some choices.

In the current world of social media there are now many forms of communication available. So, I started a blog www.lifestyledental.co.uk/blog and also a series of videos through which I began to communicate and answer some questions. Besides, I tried to explain procedures in a simple and clear way. These videos then led to this book which you are reading now and comprises of further information to help explain dentistry and dental treatment.

I am constantly learning about new dental techniques and improving in the field of dentistry so that I can provide my patients with the highest standard of care. As a dentist, it is my responsibility to diagnose, prevent and treat diseases and conditions of the oral cavity. This is not a responsibility that I take lightly, and I make every effort, along with my staff, to provide the best treatment possible to every patient. In addition to preventative oral care along with standard procedures, Lifestyle Dental Clinic also specializes in cosmetic procedures such as teeth whitening and veneers, making my practice able to address all your smile needs.

Studies have shown that oral health impacts your overall health greatly; this is a subject that everyone should be informed about. Although preventative care and treatment is the most important part of dentistry, you don't have to visit your dentist only when there is a problem or when you need a cleaning; in fact, there are a number of cosmetic procedures that you can undergo if you feel your smile needs that extra step.

My goal, along with Lifestyle Dental Clinic, is to improve the lives of our patients by helping them maintain a healthy smile while also boosting their confidence. So please enjoy my book and expect to learn a number of things that you never knew about achieving and maintaining a healthy and beautiful smile.

CHAPTER 1

The Different Treatment Options

CHAPTER 1

THE DIFFERENT TREATMENT OPTIONS

The first step to assuring a beautiful smile is by making sure your mouth is healthy. Taking care of your mouth now can ensure a healthy mouth for years to come. Your parents probably taught you to brush your teeth twice a day and floss daily but you also must visit your dentist on a regular basis so that you can nip any oral health issues in the bud. Let's review dental health issues and how they can be treated to keep your mouth healthy.

Visit Your Dentist Regularly

First of all, scheduling an appointment for a *Comprehensive Dental Examination* is your best defence of oral health problems including cavities and gum disease. At Lifestyle Dental Clinic, we take this time to thoroughly examine every part of your mouth such as teeth, gums, tongue and your lips. We will take note of the way you bite and if it has a negative effect on your teeth. We will also take x-rays of your teeth, if necessary, so we can get a better idea of any problems within the tooth itself.

In addition, we will take the time to assess any of your concerns and your goals along with any cosmetic work desired. Since this is a comprehensive exam, it could take about an hour and half to thoroughly assess your mouth and treatment options. Following this thorough assessment, we will develop a plan of treatment.

Do not wait until you are experiencing pain or discomfort before visiting your dentist. Preventative care is key, and many problems can be avoided entirely with proper dental care. Visiting your dentist twice a year for a check-up and gum therapy can catch cavities before they get too big and can also help prevent plaque build-up which can lead to gum disease. This preventative care can save you time and money by keeping your mouth healthy.

In between visits, take care of your dental hygiene by brushing your teeth twice a day. For the most thorough brushing, I recommend an Ultrasonic Toothbrush or power toothbrush. There are several brands in the market and new ones are introduced regularly. This type of toothbrush is designed to brush away plaque bacteria from your teeth and gums for a clean feeling and fresh breath. Make sure that you are also flossing every day, so that you can keep plaque from building up on your gums between your teeth, where it is difficult for a toothbrush to reach. Remember, you cannot have a healthy smile without first having healthy gums.

Many people skip these important oral health assessments because they do not have dental insurance but you should not let that stop you. Many dental clinics have their own affordable dental plans that you can purchase. In addition, most will also accept a payment plan of a monthly amount that you can afford. When it comes to your oral health, this is one expense that you can't afford to skip.

Periodontal Disease

Everyone knows that poor dental hygiene can lead to cavities but not everyone is aware of a bigger oral problem: *gum disease*. Gum disease alone should be a huge motivator to visit your dentist regularly for preventative care since many people don't even realise they have it until it is too late. In fact, gum disease is known to be a silent disease with symptoms so mild that people tend to brush it off as no big deal.

Bacteria, mucus and other particles ingested form a sticky film on the surface of the teeth called plaque. While brushing your teeth and flossing daily can remove most of the plaque, the remaining plaque will harden to form tartar. This tougher substance can be removed by a dental hygienist, and is yet another reason to visit your dentist twice a year for gum therapy.

If plaque and tartar are allowed to remain on the teeth and are not removed through proper oral hygiene, your gums will become inflamed and will begin to bleed easily. This is a mild form of gum disease called *gingivitis*. While this condition can be painful and unsightly, it typically can be corrected with gum therapy and daily brushing and flossing.

If you do not treat *gingivitis* right away, it can lead to *periodontitis* which is a more advanced form of gum disease. Once you reach this stage, your gums will begin to separate from your teeth which will form open pockets. At this point, bacteria and food particles can make their way beneath the gum line and cause the gums to become infected. Plaque

The stages of periodontal disease

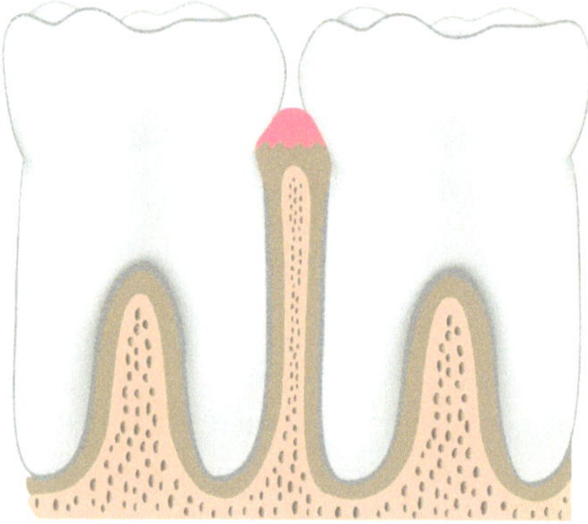

1. Healthy

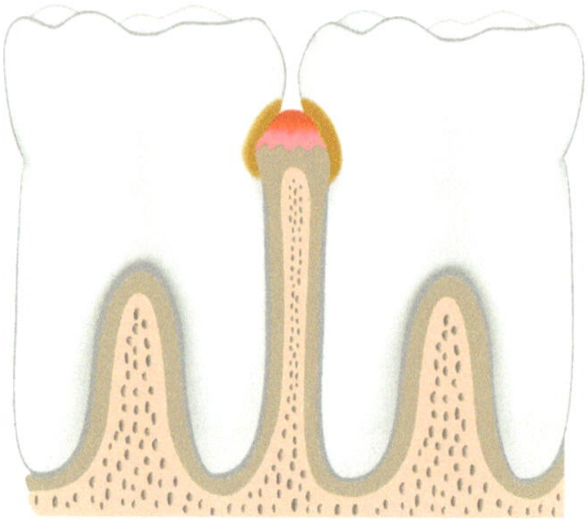

2. Gingivitis

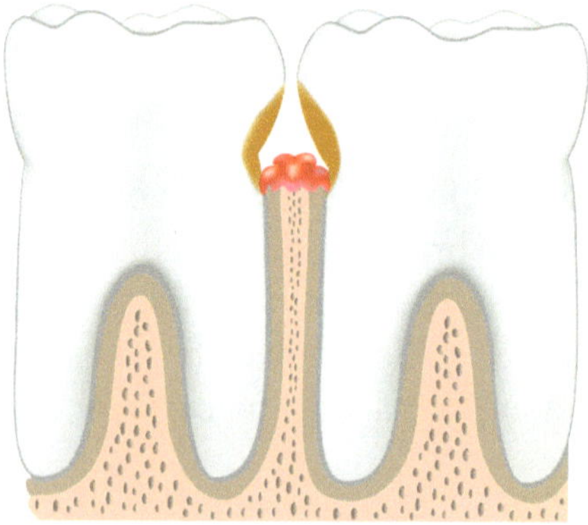

3. Periodontal pockets

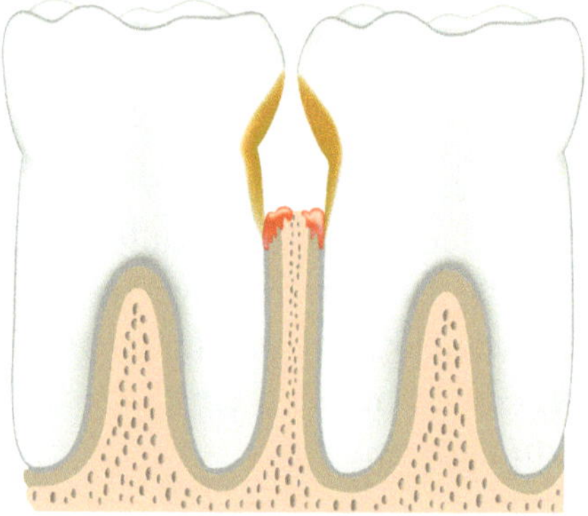

4. Periodontitis

will then spread below the gum line as well. Our immune system will naturally begin to attack the bacteria that have penetrated the surface which will cause the bone and connective tissue surrounding the teeth to break down. Over time, the supporting gum and bone will be destroyed and the teeth will become loose and will need to be removed.

There are certain risk factors associated with gum disease such as smoking, hormonal changes in women, diabetes and medications. If you have any of these factors, your gums may be more sensitive and

susceptible to gum disease so take extra precautions in keeping them clean. The first sign is typically bleeding gums so if you experience this, take care of it right away before it has a chance to advance.

To treat gum disease, you will typically receive a deep cleaning. The tartar that has built up above and below your gums will be scraped off. Then, root planing will be done to remove the bacteria from the roots of the teeth, if necessary. If the gum disease has progressed too far, surgery may be required.

If you have lost bone and tissue due to gum disease, you may need to undergo a *bone and tissue graft*. The dental surgeon may place natural or synthetic bone in order to stimulate new bone production when bone mass has been lost. They may also use synthetic or natural tissue material to cover the tooth roots that are exposed due to tissue loss.

Though gum disease can cause a great deal of damage in your mouth, it is easily preventable with proper dental hygiene. Even if you start to experience bleeding, sensitive, red or swollen gums, you can still reverse the disease effects by adopting the right oral cleaning techniques and by visiting your dentist for a cleaning. As with any health problem, catching it early always gives you the best odds of curing it with the least amount of damage.

Cavities

Cavities are holes in the surface of the teeth that are produced as a result of tooth decay. Many people mistakenly believe that only little children get cavities because of their intake of sugary foods; however, even adults are at risk of developing them. In fact, it isn't only the usual snacks such as candy, cookies, juice, cakes and sodas that cause cavities to form but also fruits, breads, cereals and milk. When we eat carbohydrates they are turned into acids by the bacteria that reside in the mouth. When this acid joins food, saliva and bacteria to create plaque

The stages of tooth decay

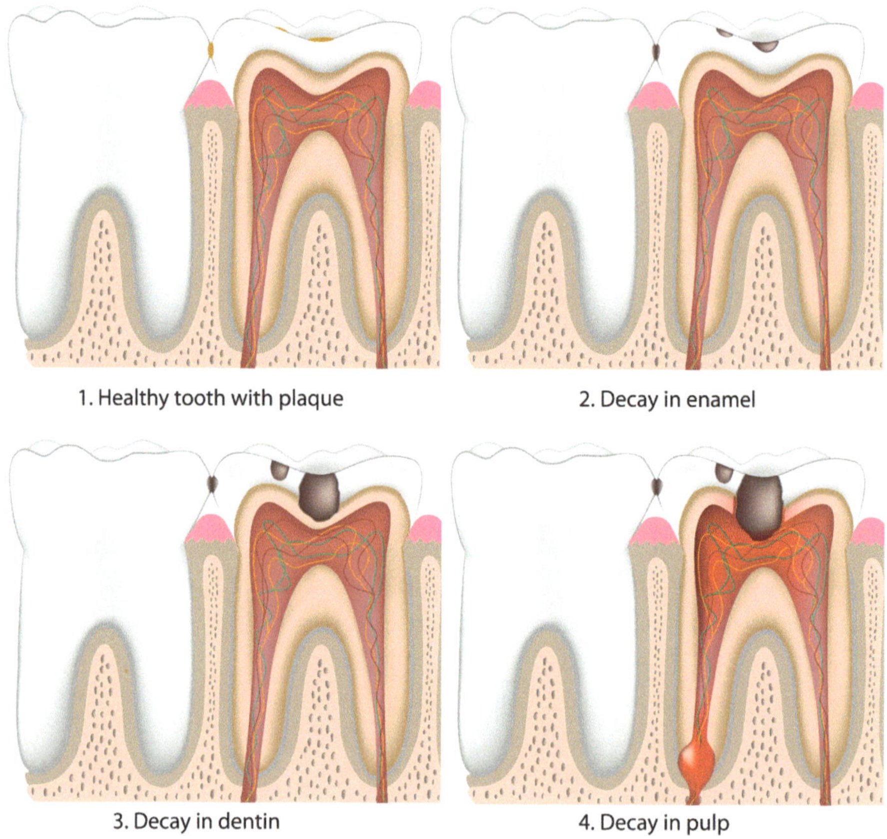

1. Healthy tooth with plaque

2. Decay in enamel

3. Decay in dentin

4. Decay in pulp

which coats the teeth, the acids will then begin to dissolve the enamel of the tooth's surface.

You may notice that you have a cavity because you are experiencing a toothache or have pain after eating sweet, cold or hot foods. You may even be able to see the pits or holes on the surface of your tooth if you are in the advanced stages of tooth decay. However, much like gum disease, if you are in the beginning stages of developing a cavity you may not even realise it. A dentist can easily identify a cavity during a check-up. He may be able to see the hole or feel it with a dental instrument. In fact, the tooth surface of a decaying tooth feels soft when being checked. Dental x-rays can point out a cavity before it is even visibly noticeable, so as long as you are receiving preventative dental care, you can catch tooth decay before it damages too much of your tooth.

Depending on how advanced the tooth decay is, there are a few different treatment options. If the decay is in its early stages, the dentist can simply fill the cavity. What this entails is removing the decayed portion of the tooth by drilling into it and replacing it with a filling made of safe materials such as gold, silver alloy, a composite resin or porcelain. If the tooth has too much decay to be removed and filled, the dentist will remove the decayed portion of the tooth then a dental crown will be put onto the tooth. Crowns are typically composed of gold or porcelain.

If you do not treat your cavity, it will continue to grow until your entire tooth is decayed. Besides the toothache and discomfort that you will experience from this, you will also lose the tooth. The whole tooth will need to be extracted and an implant put in its place. Not only can this be painful but it can be costly, too. Visiting your dentist for a check-up every 6 months can prevent a cavity from reaching this stage of tooth decay.

Root Canals

While tooth decay is a leading cause for needing a root canal, other reasons include repeated dental work on the tooth, a large filling in the tooth, a crack or chip in the tooth or some sort of trauma to your mouth. These occurrences damage the nerve or pulp of the tooth and cause it to die, thus making a root canal procedure necessary. During this treatment, the centre of the tooth is extracted, such as the nerve, blood vessel and tissue, and the area is cleaned out. The decayed parts of the tooth will also be removed. The root of the tooth is then filled in with sealing material and a dental crown is put on to cover the

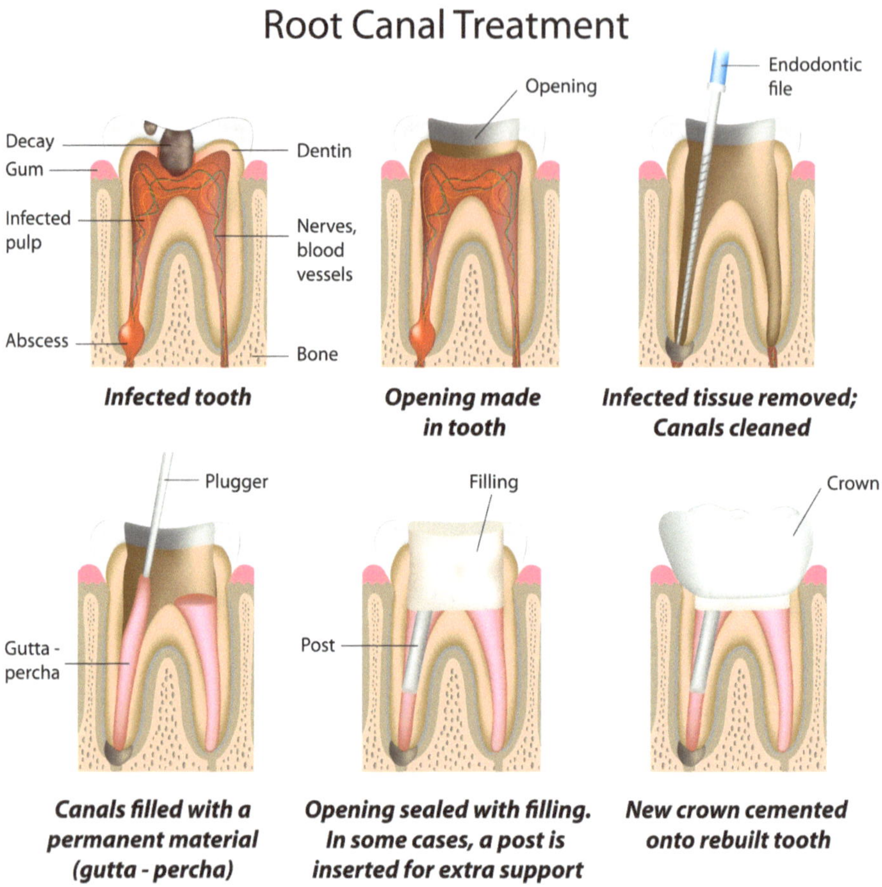

Root Canal Treatment

Opening

Endodontic file

Decay

Dentin

Gum

Infected pulp

Nerves, blood vessels

Abscess

Bone

Infected tooth

Opening made in tooth

Infected tissue removed; Canals cleaned

Plugger

Filling

Crown

Gutta - percha

Post

Canals filled with a permanent material (gutta - percha)

Opening sealed with filling. In some cases, a post is inserted for extra support

New crown cemented onto rebuilt tooth

tooth, if necessary. Typically, the crown is put on your tooth several days after the root canal. A root canal will reduce your risk of infection while allowing you to keep your tooth without the prior toothache.

If the nerve tissue and pulp of the tooth are left untreated, the bacteria will multiply causing infection. If the infection spreads down to the root of the tooth, an abscess could grow. An abscess is a pocket filled with pus at the root of the damaged tooth and can cause you more pain which can radiate to your jaw. An infection of your root

canal could also cause your face, neck and head to swell; loss of bone at the root's tip; and drainage leaking into your gums or cheek.

Some signs that a root canal is necessary include tooth pain when chewing food or applying pressure, sensitivity to hot and cold foods long after taken out of your mouth, darkening of a tooth, swelling or tenderness of the gums around the tooth, and a pimple on the gums.

Many people believe that the pain of a root canal is great and therefore put off having it done. The truth is the pain that you will experience from the decaying tooth and nerve damage is far worse than the pain of having the root canal done. In fact, with anaesthesia and ever-improving dental technology, a root canal procedure can be performed with little, if any, pain at all. Following the root canal, you may have some sensitivity due to tissue inflammation. This can be treated with over the counter pain relievers. You can typically even return to work the very next day after the procedure.

Tooth Extraction

There are many reasons that can arise which make it necessary to have a tooth pulled out of your mouth. One major reason is allowing tooth decay or gum disease to destroy your tooth to the point that it needs to be pulled out. This is done to prevent infection from spreading to other teeth and parts of your mouth, as well as relieving pain. Wisdom teeth are a common extraction seen in younger people. When there is no room in the mouth for the wisdom teeth to come in or they are not coming in correctly, the wisdom teeth are often removed – sometimes even before they break through the surface of the gums. Another reason for having a tooth pulled is if your mouth is over-crowded and you are being prepared for orthodontic braces.

The dentist will typically apply a local anaesthetic to the area around the tooth to be extracted. If more than one tooth is to be extracted or if

13

removing an impacted tooth, the dentist may choose to administer general anaesthesia which will cause you to sleep throughout the procedure. The dentist will then grasp the tooth with forceps and move it from front to back (similar to how kids try to loosen a baby tooth) in order to remove it from the jaw bone and surrounding ligaments. If the tooth is impacted, the dentist will first cut away the gum and bone tissue above the tooth. Sometimes when a tooth is hard to pull, it must be removed in pieces.

Depending on the situation, the dentist may choose to close the gums of the extraction site with a couple of self-dissolving stitches. The site will be covered with gauze and the procedure will be complete. A blood clot will naturally form in the empty socket to stop the bleeding and your gums should heal within a couple of days. Again, over the counter pain relievers should be enough to treat any discomfort that may result from the extraction.

If the tooth was extracted due to damage, you may choose to have an implant, denture or bridge put in the now empty socket. A dental implant is a titanium post that is inserted into the bone of the socket. You can then have a crown or bridge put on top of it to stimulate the look of having a real tooth there. As long as you have enough bone, you should be able to have a dental implant put in.

COMMON DENTAL ISSUES

Sensitive Teeth

There are many reasons why you could be experiencing tooth sensitivity. One could be that you have tooth decay due to an untreated cavity or gum disease. Decayed teeth can leave the nerve exposed which causes sensitivity when touched. If your gums are receding or you have a crack in your tooth, you could also experience tooth sensitivity. A filling may be all you need to cover the exposed nerve and decrease your tooth's sensitivity.

Teeth wearing away

There are a few main reasons for your teeth to wear away. One is because of grinding. Many people get into a subconscious habit, especially at night, of grinding their teeth. If you find that your jaw is sore in the morning or you catch yourself grinding your teeth during the daytime as a response to stress, then grinding is probably to blame for your teeth wearing away. Another reason can be tooth erosion from acid. The main culprits in eroding your enamel are soft drinks and acidic juices such as orange juice. Try lowering your consumption of these kinds of drinks and brushing your teeth immediately after drinking. A final reason could be due to excessive brushing. While brushing your teeth is necessary for good oral health and keeping your teeth clean, there is such a thing as over brushing. Twice daily should be enough to do the job. In addition, use moderate pressure when brushing your teeth and avoid brushing too hard as this can thin your enamel down, too.

Jaw Pain

If you experience jaw pain not associated with tooth decay, it could be because of the way you bite. If you have an over bite or an under bite, it can cause your top and bottom jaw to be uneven thus causing discomfort. To fix this, a bite plate or retainer can be custom made by your dentist for you to wear. Over time, your bite will be corrected and your jaw pain will be relieved. Another big reason for jaw pain is from clenching your teeth. This is when you tense your jaw and keep your teeth closed tightly together. Many people tend to do this as a response to stress and often don't realise they are doing it until their jaw becomes sore.

Fear of visiting the dentist

There are many people that have a very real fear of visiting the dentist simply because they are afraid of the pain of dental procedures.

If this is the case for you, you can try using sedation for a dental procedure. Sedation will help you relax so you are not tense or anxious about the impending treatment. One popular form of sedation is administering medazalom through an IV into the patient. This medicine has the effect of alcohol and will relax the patient completely so they can undergo treatment without any anxiety.

CHAPTER 2

Personalize Your Treatment

CHAPTER 2

PERSONALIZE YOUR TREATMENT

While the basics of dental health remain the same for everyone, there are different options for every person. Depending on your lifestyle, health and even preferences, there is a different approach and course of treatment for you.

Busy People and Professionals

You may feel like you are too busy with work and your children to make time for the dentist but taking care of your mouth can be an investment in your future. First of all, if you don't make time for a one hour cleaning and check-up twice a year, it can result in even more lost time later on. Unbeknownst to you, you could have tooth decay or gum disease (or even both!) and without twice yearly check-ups, they could grow into advance stages. Eventually, when the pain is so severe, you will finally need to make time for the dentist; but at this point, it would cost you even more time to fix the damage and also more money out of your pocket. So no matter how busy you are, you need to make time for dental check-ups to save you time in the future.

Another reason to get in to see your dentist is that having a healthy smile can benefit your career. You have probably heard the expression "a million dollar smile" in relation to successful movie stars. Do you think they would be quite so successful without it? The truth is that your smile is one of the first things that a person notices about you and your boss or potential employer is no different. In order to make the best impression possible, you need to take care of your mouth. As superficial as it may seem, your smile can make or break you.

A smile can be contagious; if you are smiling, you send out positive vibes that those around you will pick up on. Therefore, if you are not smiling, you won't send out those positive vibes. When you are interviewing for a new job or trying to prove yourself to your boss for a promotion, a smile can help you get it. The reason many people don't smile or have a tight lipped smile is because they are dissatisfied with their teeth.

Some people may say that your physical appearance doesn't matter when it comes to your career, but the truth is that how you present yourself does. Whenever you meet someone, whether it is someone interviewing you for a job or someone asking you out on a date, they will form a first impression of you almost instantly. This is why when you first introduce yourself and shake their hand, you should give them a warm, friendly smile. Without it, they can quickly get the wrong impression of you and this can take a long time to be reversed.

By visiting your dentist regularly, you can keep your teeth and gums looking good. Tooth decay and gum disease can leave your teeth discoloured and your gums read and swollen. It can also erode away your teeth and gums, leaving you with an unhealthy looking smile. By taking care of your mouth, you can ensure that your gums and teeth look healthy and will give you the confidence you need to put your best foot forward and smile.

Diabetics

When you have diabetes, you have to take extra care of your health, and your dental health is no different. In fact, just by taking care of your oral health you can detect if you have developed diabetes sooner than you normally would. Your mouth is typically the first indication

you have of your overall health. Many diseases and health problems, such as diabetes produce oral signs and symptoms.

In addition, if you have diabetes, some dental conditions can make it even worse. For example, *periodontitis* can make it difficult for diabetics to control their blood sugar level. Diabetes also leaves you at a bigger risk of oral health issues because of the fact that it is difficult to control blood sugar. The bigger problem you have with controlling your blood sugar, the bigger risk of dental problems. This is based on the fact that uncontrolled diabetes can damage your white blood cells whose purpose is to fight off bacterial infections. This leaves fewer white blood cells to fight off oral infections which allow the infection to continue to spread. Uncontrolled diabetes also presents the risk of more cavities since saliva will contain more sugar than usual. This higher level of sugar can help bacteria thrive and can lead to the development of tooth decay.

If you are a diabetic, you are more prone to *gingivitis* and *periodontitis* and tooth decay. In fact, if you are a smoker then your chances of developing these conditions are 20 times increased. This is mostly due to the fact that smoking can limit blood flow to your gums which can prevent or prolong healing. Visiting your dentist twice a year for check-ups as well as visiting whenever you are having any pain can help to keep these dental conditions at bay and to diagnose them before they become a real problem.

Whenever you go to your dentist for an oral health assessment, be sure to inform them of your current status. For example, what your *HgA1C* level or when you took your last dose of insulin. If you are going to be treated for periodontal disease or have oral surgery, consult both your dentist and your diabetes doctor beforehand. They may recommend that you take pre-surgery antibiotics or that you change your meal or insulin schedule. Always follow any post treatment instructions that your dentist may give you since diabetes will make the healing process longer for you.

As long as you are diligent with brushing twice a day, flossing daily and visiting your dentist twice a year and also control your sugar and insulin levels, having diabetes doesn't have to keep you from having a healthy mouth.

Pregnant Women

When a woman becomes pregnant, it is so easy to become consumed with prenatal health that dental health is sometimes put on the back burner. However, your teeth and gums cannot be ignored – not even if it's just for 9 months. Because of the hormonal changes that pregnancy brings on, your mouth goes through changes just as the rest of your body does. In fact, many pregnant women are at a higher risk of developing *gingivitis* and *periodontitis* which can actually affect the health and development of your growing fetus.

You can decrease the risk of your baby being negatively affected by your dental problems by giving your mouth as much attention as the rest of your body. Like always, prevention is key here so if you are planning on becoming pregnant you should schedule a dental appointment for a cleaning and oral health assessment to determine if there are any oral problems. As a matter of fact, even if you aren't planning on becoming pregnant any time soon but still have the possibility of becoming pregnant, you should also ensure you visit your dentist every 6 months.

When you become pregnant, you should keep up with your dental appointments as usual and let your dentist know that you are pregnant at your next appointment. Typically, dental treatments are not performed during the beginning and end of your pregnancy unless absolutely necessary. This is because the first trimester and the last part of the third trimester are critical times of development for your baby so dentists and doctors usually refrain performing any non-essential work at this time. You regular dental check-ups and cleanings should be scheduled to fall into the second trimester of your pregnancy since there is

22

no major growth that is done during these months besides weight gain. Any elective dental treatments should be done after the baby is born. Your yearly dental x-rays will also be postponed until after the baby has been delivered except in an emergency case. X-rays are much safer than they used to be; however, doctors and dentists take every precaution to protect the health of your unborn baby. If you have been an accident or have another dental emergency then a dental X-ray will be performed. If it isn't urgent, then the x-ray will be scheduled for after your baby's birth.

Since pregnancy makes you more prone to tender, sensitive gums, you should pay extra attention to them. Continue to floss in order to prevent tartar build up since pregnant women are at an increased risk of developing periodontal disease. If your gums start to bleed or become red and swollen, make an appointment immediately with your dentist because you may have *gingivitis*. You want to treat it before it reaches an advanced stage of *periodontitis* because this condition increases a woman's chance of delivering a premature or low weight baby. The reason for this is believed to be because the bacteria in the mouth release toxins that can inhibit the growth of the fetus. The bacteria can cause an infection which can cause the woman to produce hormones that induce labor. This can then lead to premature birth and/or a baby with a low birth weight.

Many pregnant women experience morning sickness throughout a portion of their pregnancy which creates two problems: one, they may skip brushing their teeth because the toothpaste makes them sick; and, two, the stomach acids brought up when they vomit can damage the teeth. To combat the first problem, try switching to a bland toothpaste and avoid flavoured ones that may make you sick. To help with the second problem, when you do get sick, be sure to rinse your mouth out well immediately afterward to wash away any acid or food.

After you have the baby, you may get so caught up with the needs of your baby that you neglect your own health. Be sure to keep up all your

23

good dental habits not only for your health but also so you can then pass them on to your child.

Fear of the Dentist

If you are nervous or scared about going to the dentist, you are not alone. There are a large percentage of people that avoid going to the dentist because of their fears whether it is their fear of pain, that the dentist will find something wrong or because they are just afraid of doctor offices in general. Because of this, these people will allow a small dental problem to advance into a big problem, and since dental health can translate into other parts of your health, you could end up with a serious problem.

The good news is that there are different techniques that can be utilized to ease your anxiety over going to the dentist or having dental work done. The sedation technique is one that many dental clinics utilize on a regular basis. It involves injecting a sedative intravenously which means a needle will placed directly into your vein and will pump in the correct dosage. This technique will cause you to feel relaxed and comfortable while your dentist performs dental work.

Another method is hypnosis. A 20 minute hypnosis session is usually adequate for the patient to relax enough for the treatment to be done. Hypnosis involves reaching the patient's subconscious state and telling them that they will not experience any nervousness or fear when they have the dental work done. When the patient is brought back to consciousness, they then will not have any fear or anxiety about it.

We also take the extra step at *Lifestyle Dental Clinic* to assure our nervous patients. We use something coined a *"magic buzzer"* that the patient can hold onto during treatment. If at any time they feel uncomfortable they can press the button and our staff will stop treatment immediately. We will then take the time to address the

problem and determine a way to get past it. This is a great tool because it allows the patient to feel in control of their treatment at all times. In addition, we play calm, quiet music in the dental exam room which can help put your mind at ease and allow you to rest.

Those Focused on Total Health

As I have mentioned before, your dental health is a good indicator of your overall health. If you have an unhealthy mouth then you are prone to many other health conditions. The biggest reason is that bacteria from oral infections can travel through your bloodstream and cause new infections in other areas. Your whole body works together in one unit, so an illness in one area can easily spread elsewhere.

Dental researchers have discovered that there is a direct link between *periodontitis* and cardiovascular disease. This is primarily because the bacteria in your mouth cause inflammation of your arteries throughout your entire body. *Gingivitis* in particular can contribute to clogged arteries and even blood clots. Both of these conditions can lead to heart attack or stroke which can leave you incapacitated or can even lead to death.

Oral infections can also negatively impact your respiratory system. Oral bacteria can be inhaled and deposited into the lungs which could cause lung disease in someone who is already battling an illness and has a lowered immune system. These bacteria can also advance an already occurring case of lung disease and cause respiratory distress. It can even increase your risk of developing pneumonia by three to six times.

In addition to your physical health, your dental health can also affect your emotional health – particularly your love life. There is nothing worse than bad breath when trying to talk to someone you are romantically interested in. Bad breath has two effects: one, it lowers your confidence; and two, it can keep people away from you. If you know you have bad breath, it can keep you from approaching someone that you are interested in. Likewise, if someone smells your bad breath, it is likely that they will make up an excuse to get away from you. Bad breath is not attractive and can really curb your love life.

By brushing your teeth twice a day and flossing, you should be able to prevent bad breath. Rinsing out your mouth with water or mouth wash after a meal can also help to get any remnants of your food out of your mouth. If you find that no matter how clean you keep your mouth you still have bad breath or you have particularly foul breath then make an appointment to visit your dentist. It is possible that you have a condition called halitosis. Halitosis is typically related to a bacteria associated with gum disease. When this certain type of bacteria feeds on food particles, it produces a foul odor in the process. This condition is usually a result of poor oral hygiene, but by brushing your teeth and gums in addition to flossing, you can prevent it from developing.

If You Have a Lot of Dental Work

If you have a lot of dental work to be done, your dentist will work with you to get it done with a plan that will not cause you too much discomfort or too much financial distress. If you neglected your teeth when you were younger or if you were in an accident or even suffered from an illness, you may need a great deal of treatment to get your mouth healthy. Perhaps you have multiple cavities to be filled, a root canal that needs to be done and a deep cleaning and dental hygiene to reverse the effects of periodontal disease. No matter what you need to get done, or how much you need to get done, you can work out a plan that suits you best.

You don't have to worry that you will need to spend an entire day at the dentist getting every treatment at once because your dentist will not do that to you. Likely, your treatment will be split between a couple cavities being filled one day, a couple more on another day, the root canal at

26

another time and the deep cleaning and dental hygiene on yet a different day. Extensive dental work like this is spread out over the course of many weeks so that there is not too much trauma to your mouth at once. Likewise, this will spread the payments out over a period of time so it isn't too overwhelming. Most dental clinics offer payment plans that allow you to pay a monthly amount that you can afford which is a great help.

Your dentist will help you decide which dental treatments are priorities and which can be pushed back to the end of your course of treatment. If you are a new dental patient, be sure to let your dentist know in advance what kind of dental work you have had done and any dental conditions you have ever had. This will allow your dentist to provide the best care individual to your mouth.

No matter who you are, what situation you are in and what kind of dental treatments you need, your dentist can create a personal dental plan to properly suit your individual needs. Once you have your oral health taken care of, you can decide if your mouth needs a makeover and what type of cosmetic dental services are right for you.

CHAPTER 3

Cosmetic Dentistry

Chapter 3

Cosmetic Dentistry

Once you have a healthy smile, it is time to get a beautiful smile. Since your mouth is the first thing that people notice about you, it is important to have a smile that you are proud of. Luckily, there are many cosmetic treatments that can help your smile shine.

Teeth Whitening

One of the biggest complaints that patients have about their teeth is that they have a yellowing, dull colour. One of the simplest and most cost effective cosmetic treatments that you can have done is getting your teeth professionally whitened. Just by changing the shade of your teeth, you can make a dramatic difference to the appearance of your smile.

There are many foods and drinks that you consume that will contribute to the staining of your teeth. Over time, you will notice your teeth turning yellow or brown which cannot only be unattractive but embarrassing, too. Dark drinks in particular cause the most staining. Coffee and red wine are leading culprits in staining teeth so if you can't reduce your consumption of them, then you should at least rinse your mouth out with water afterwards.

Staining in itself poses no true dental risk, hence the reason that teeth whitening is considered cosmetic. However, stained teeth point out the fact that your teeth are covered with plaque and that you may not be keeping up with your dental hygiene. When plaque builds up and hardens, it causes tartar. Tartar is stained much easier than healthy tooth enamel. So if you notice discolouration of your teeth, you may not only need your teeth professionally whitened, but you will also

need gum therapy. Since tartar can lead to gum disease, you need to take care of this issue to ensure that your mouth will be healthy below those pearly whites.

Are Coffee and Wine to Blame?

Besides discolouration, coffee and wine can also cause some real damage to your teeth. Because of coffee's acidity; calcium and phosphate can erode right out of your tooth enamel. If you use sugar in your coffee then you also risk tooth decay. Wine is equally destructive. Everyone knows that red wine stains your teeth but not many people realise that white wine is damaging, too. Studies have shown that consumption of white wine contributes to rapid tooth erosion. You can help protect your teeth when you drink these beverages while you are eating.

When you eat, you produce more saliva which fights enamel erosion. It is a good idea to rinse your mouth out with water after drinking an acidic beverage but refrain from brushing your teeth too soon because doing so can actually damage the tooth's structure. Instead, brush your teeth about an hour after you have consumed such drinks.

As long as you take care of your teeth, you don't have to completely eliminate the drinks you love. Regular brushing, flossing and dental examinations can prevent much damage being done to your teeth.

Is Professional Tray Whitening a Good Option?

In response to the recent peak in interest of teeth whitening, many companies have tried to cash in by creating their own teeth whitening system that consumers can buy in stores. They market these products as a cheaper way of receiving professional results that you can achieve in the comfort of your own home. However, the age old adage *"you get what you pay for"* rings especially true here.

32

Some people are able to maintain their shining smile by being diligent with their dental care but for most people, they need that extra step to get their smile the way they want it to be. This is when you should turn to your dentist and not to a box in your local chemist's shop. The main

difference between your dentist and a boxed whitening kit at the store is that your dentist has gone through years of schooling, training and hands-on experience to whiten teeth whereas the box has not.

Not everyone is a candidate for teeth whitening and some people have sensitivities that must be handled differently. Your dentist knows this; a box does not. Just look at the over the counter options and see for yourself:

Whitening Toothpaste

This comes in a typical toothpaste tube and contains about 10% of peroxide. The directions say to brush the paste on with your

toothpaste two times a day and wait five minutes before rinsing out your mouth. This may seem like a quick, easy and cheap option but you can quickly damage your tooth enamel and your gums. It can also leave your teeth extra sensitive.

Whitening Kit

These are the most common teeth whitening products on the market right now and you probably can't go an hour watching television without seeing at least one commercial for them. They come in a few different forms such as trays and strips at a reasonable price but again, at a high risk. The acidic gel used can erode your tooth enamel and even your dental work. Because it contains higher amounts of peroxide (10-16%) it can also make your teeth very sensitive to temperature that can last for weeks. Also, if you accidentally swallow any of this solution, it can make you very sick.

Professional Whitening

Your dentist will use the most effective and safest way to whiten your teeth depending on your dental history. If you have stubborn or dark stains on your teeth, your dentist may recommend power whitening. This treatment is done in the dental office and under the dental team's watchful eye. A gel consisting of 35% peroxide is put on your teeth and then a special light will be aimed at your teeth for about two hours to quicken the whitening process. To protect your gums and lips, a gum guard or dam will be placed in your mouth. If a two hour long session doesn't suit you, the time can be split up over the course of a few sessions.

However, if you have receded gums or cracked enamel or dentin, this may not be the best option for you. This high concentration of peroxide and long exposure can irritate the skin between your teeth and make your teeth extra sensitive. In this case, your dentist will determine if there is a better whitening option for you.

Another common treatment is a prescription whitening kit. With this option, your dentist will create a mouth tray that is designed to fit your mouth perfectly. The peroxide gel is much thicker and stickier than over the counter versions so you will be less likely to accidentally swallow any. Your dentist will show you how to properly use it so that you can bleach your teeth on your own time and in the comfort of your own home. You will then see your dentist about a week later to ensure that your teeth, gums and dental work are not being damaged by the peroxide.

If your teeth develop a sensitivity to hot and cold foods and drinks while using the whitening tray, let your dentist know and he can prescribe a fluoride gel that is applied after whitening to ease it. However, a majority of people don't have much sensitivity to the process at all.

After you complete the whitening process set forth by your dentist, the effects should last about five to six years. Depending on your diet, especially your intake of drinks such as coffee and wine, it may not last as long. If you maintain good dental habits and hygiene, your teeth could remain white for upwards of six years. At *Lifestyle Dental Clinic,* we will go over what types of foods, drinks and activities will shorten the lasting effects of teeth whitening. We will also discuss what you should do keep your teeth as white as possible.

Orthodontics and Braces

If you are unhappy with your crooked or misaligned teeth, braces can straighten them out and give you a smile to be proud of. Your dentist or orthodontist will take impressions of your teeth and have x-rays done. Once this is done and they can see exactly how your teeth need to be aligned, they will develop an orthodontic plan for you. Most people can be treated with braces; however, if you only need a slight realignment you may just need to wear a retainer; and if you have serious misalignment, such as extreme under bite or overbite, you may need surgery.

There are different forms of braces available these days but the most common are the traditional wire and brackets and invisible, removable ones. Braces straighten teeth slowly over time by pushing them in the direction that they should be in. While everyone's mouth is different, braces are usually kept on anywhere from 1 to 3 years. Once the braces are removed, a retainer needs to be worn around the clock for about six months. After that, it should be worn while you sleep for a year or more depending on your dentist's recommendation.

While having orthodontic treatment, you will visit your orthodontist on average every month or six weeks to ensure that your braces are doing their job. At this time, he or she will make adjustments to the wires, bands and springs to cause more tension and move the teeth even more. After these adjustments, your mouth will typically feel sore and uncomfortable for the following few days. Over the counter pain relievers and eating soft foods can help ease this soreness.

After your orthodontist sees that your teeth have been completely straightened and aligned, the braces will be removed. New x-rays and bite compressions will be done to compare to the ones from before treatment started. You will then be fitted for a retainer which is custom made to fit every impression of your teeth. This is used to correct any shifts in the placement of your teeth.

Orthodontic treatments are typically seen in pre-teenage children but the benefits are the same for everyone. It is never too late to get braces, and more and more adults are having orthodontic treatment these days as a way of repairing their smile. A retainer may need to be worn for a couple of years before the bone, gums and muscles surrounding your teeth adapt to their new placement. As time goes on, teeth tend to shift a little so using your retainer can bring them back to their correct positions.

Six Month Smile

If you need your teeth straightened but don't want to wait the long time that traditional braces take, we have a brace system that is complete in just 6 months. This is just half or even a quarter of the time that traditional braces would stay on for. Most people are good candidates for this treatment but there are a few who for one reason or another will need their braces to stay on a bit longer. Just imagine: having braces put on at one oral health assessment and then by your next oral health assessment in 6 months, your braces come off!

After your treatment and all your teeth are perfectly aligned, you may find that they may not be level on the top. This is any easy fix; by just contouring the teeth, the tops can be smoothed out or built up depending on the situation. Cosmetic contouring is a great way to get your teeth looking perfect.

Smile Makeover

The Smile Makeover at *Lifestyle Dental Clinic* is the ultimate cosmetic dental package. First, to determine which treatments are best for your makeover, we will sit down and discuss your options. We will ask what you don't like about your smile and what we can do to fix it. After we ascertain this information, we will use prototypes to create a model that will show what your smile will look like after the makeover. This way, you know up front what your smile will look like and if are happy with its appearance. At this point, any changes to the planned makeover can be made until you are fully satisfied with the anticipated appearance.

Typically, the makeover will involve the placement of 10 or more veneers or crowns on the top and bottom teeth in order to give your smile a better shape and a lighter appearance. Since you are having multiple veneers or crowns placed in, there is a discount on the total cost, making this a cost effective option. Afterward, you will have a mega-watt smile that people won't be able to do anything but notice. This is a great treatment to have done for a milestone birthday since a great smile can visibly knock 10 years off of your age, so go ahead and treat yourself!

Crowns

A crown is a cap that is placed over a tooth that has lost size and strength due to tooth decay, gum disease or injury. When the crown is anchored in place, it will cover any visible portion of your remaining tooth right up to the gum line. Dental crowns can be made of many different materials such as metals, porcelain fused to metal, all resin, ceramic or porcelain.

Having a crown put in is a two-step process typically spread over two office visits. During the first visit, your dentist will take x-rays of your tooth root and surrounding bone. If there is extensive tooth decay, you may need to have a root canal first. Your dentist will administer a local anaesthesia of the surrounding area before filing the tooth down. If too much of your existing tooth is damaged due to decay, it may need to be built up enough for the crown to be supported. Next, an impression will be made using putty in order to have the crown made. It typically takes 2-3 weeks for a dental crown to be made in a dental laboratory and returned to your dentist's office. Until this is received, you will have a temporary crown held in place with temporary cement.

During your second visit of the process, the temporary crown will be removed and the permanent crown will be put in using permanent cement. You may experience soreness after the treatment but this is normal and can be relieved by using over the counter pain relievers. Crowns typically last for 5-15 years on average. If you maintain good oral hygiene and refrain from grinding, clenching or using your teeth to open things, they will last much longer.

Cosmetically, porcelain crowns provide the most natural look. If you have old metal crowns, consider having them replaced with porcelain for an immediate mouth brightener. The old crowns are easily detected by others because of the black line that is shown at your gum line. However, porcelain crowns are completely natural looking so no one will ever know that they aren't your own teeth. By simply replacing these crowns, your smile will have an instant rejuvenation. Even though a crown is artificial, you still need to brush it and floss between it like you would a regular tooth. The natural part of your tooth is still subject to tooth decay as is the underlying gum to gum disease.

Bridges

Just like a bridge over water, a dental bridge brings two sides together. If you have a missing tooth and want to bridge the gap, it can give you more confidence in your smile. Similar to crowns; bridges are also a two-step process. During the first visit, your dentist will re contour the teeth on either side of the gap so they can be fitted for crowns. Impressions will be made to create the crowns, pontic and the bridge. While you wait for these to be created in the dental laboratory, you will have temporary versions placed to protect the area.

During the second visit, the crowns will be put on each tooth next to the gap in order to anchor the bridge. Then the bridge will be permanently cemented in place. Sometimes a dentist will only use temporary cement for the bridge and wait a few weeks to ensure it is a proper fit before permanently affixing it. Like crowns, bridges typically last 5-15 years but with good oral care they can last even longer. Once you are adjusted to your bridge, not only will your smile be improved but you will find it easier to eat without the gap in your mouth. Likewise, you may find that your speech has improved as well. When you are all done, you will be able to smile wide again.

Dental Implants

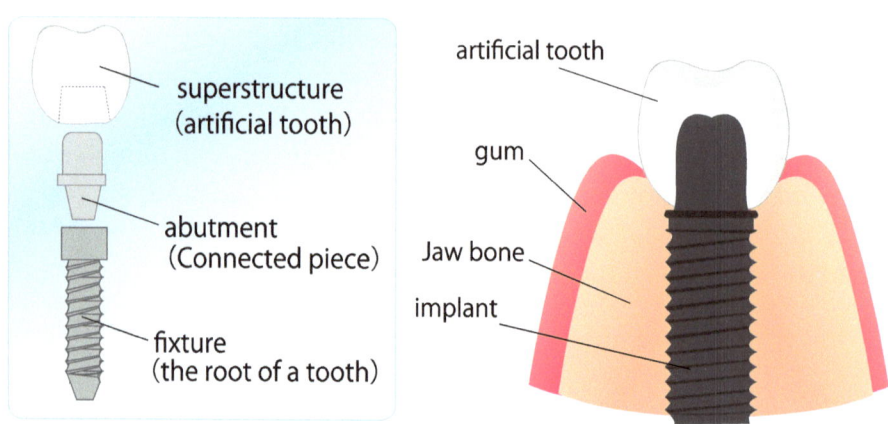

superstructure
(artificial tooth)

abutment
(Connected piece)

fixture
(the root of a tooth)

artificial tooth

gum

Jaw bone

implant

If you have lost teeth due to tooth decay, gum disease or injury and want to fill in the gaps of your smile, dental implants can give you a permanent solution. No more gaps in your smile or having to take your dentures out. The first step is the tooth root implant which is a titanium post which is placed directly into your bone in the socket of where the tooth will be placed. It can take several weeks for the bone to heal and grow around the post. Once this happens, the post will be secured in the jaw. Next, an abutment which is a small connector post will be attached to the post bonded to the bone. This is used to secure the crown in place. The crown is created by taking impressions of your existing teeth so the dentist can create a new one that would naturally fit in with the rest. He or she will also be sure to have the crown be the same shade as the rest of your teeth. The goal here is to make the implant as natural as possible so that it looks like every other tooth in your mouth.

In order to be a good candidate for dental implants, you need to have healthy gums and enough existing bone that can hold the implant in place. Therefore, any gum disease should be addressed prior to having

dental implants put in. The pain of the procedure is dulled with local anaesthesia and shouldn't hurt more than a routine tooth extraction. Any gum soreness can be treated with over the counter pain relievers.

The benefits of dental implants are plenty including a more natural appearance; improved speech; improved comfort; easier to eat; better oral health; durable; convenient and self-esteem booster. Since they are permanent, they don't have to be taken out and soaked to be cleaned. You can care for your dental implants just as you would you real teeth with regular brushing, flossing and twice yearly gum therapy and oral health assessments.

Veneers

Veneers are porcelain or resin shells that are place on the front of the tooth's surface to cosmetically change their appearance. When bonded on, they change the colour and shape of your teeth to create a better smile. Dental veneers are a great option if your teeth are discoloured, worn down, chipped, broken, misaligned or uneven, or have gaps between them.

Having veneers put on is typically a three visit process. The first visit is the consultation where your dentist will discuss if veneers are right for you and possibly take x-rays and impressions done. At the second visit, your dentist will need to remove a small portion of your enamel. This is usually about ½ millimeter of the tooth's surface. After this is done, you will have impressions done so that the dental lab can make your veneers as a custom fit. After a week or two, you will come in for your third appointment where the veneers will be bonded to your teeth. But first, your dentist will place the veneers on your teeth to ensure that they are a perfect fit and colour. The veneers can be trimmed down a bit if necessary and the colour can be changed with the cement. Once the dentist is sure the veneers are right for your teeth, he or she will

clean, polish and then etch them to allow easier bonding. The veneers are then permanently bonded to your teeth with a special cement and then a special light will help it dry quickly. Your dentist will next clean off cement from your teeth and make little changes until perfect.

Your dentist may advise that you steer clear of dark liquids such as coffee and red wine to avoid staining of your veneers. Veneers typically last anywhere from five to ten years and then they will need to be replaced. You should continue to practice good oral care including brushing and flossing.

White Composite Fillings

These days, more and more people are choosing white composite fillings for their cavities instead of the typical amalgam fillings. In fact, many people have even had their amalgam fillings replaced with white fillings. The biggest benefit that white fillings provide is that they appear more natural. Think about it: if you have a mouth full of amalgam fillings and you open your mouth wide, people will be able to see all the fillings that you have had done. Not only is this unsightly but it is also embarrassing since this is a way of people figuring out the amount of cavities you have had. However, if your cavities are filled with white composite, no one will even notice them.

In addition to cosmetic benefit, composite fillings have other benefits over amalgam, too. Composite restores strength to the tooth while amalgam weakens it and makes them prone to breaking. Composite also can restore the tooth's natural appearance. Amalgam filled teeth are typically more sensitive to temperature than composite filled teeth. Less of the tooth needs to be removed for the composite filling than with the amalgam. Finally, perhaps the biggest benefit is that composite fillings are mercury free unlike amalgam fillings. When it comes down to it, the choice between the two is clear. For your tooth's health as well as a better smile, composite fillings are your best option.

Gum Contouring

If you have gums that seem to cover too much of your teeth or not enough, gum contouring may be a good option for you. This is often a result of having a crown or implant put in your mouth and can leave it looking unnatural. However, genetics and health problems including *periodontitis* can also be behind this. Gums that are too high can make your teeth look too small, while gums that are too low can make your teeth look too big. Sculpting your gums can improve the appearance of your smile and can help make crowns and implants look more natural like they are your real teeth.

Your dentist can sculpt your gums right in their office by using a laser or scalpel. The dentist will use a special pen to mark on your gums to show you where they will be leveled off at or built-up if needed. This way, you can see beforehand the new shape that your gums will have before the procedure is performed. Your dentist will apply local anaesthesia to your mouth in order to completely numb the area so you feel no pain.

It will take anywhere from a couple of days to a couple of weeks for your gums to fully heal. Your dentist will provide detailed instructions at the time of the procedure of what you should eat, drink and how to care for your gums while you heal. Over the counter pain relievers should be enough to ease the pain but you should avoid aspirin because it can cause bleeding. Stick to soft foods in the days following the surgery and don't eat anything warm or hot in temperature. Foods such as yogurt, soft vegetables, ice cream and cold pasta are good options. Also avoid spicy foods until your gums are completely healed as the spice can really burn in your open wounds. The risks of having a gum contouring procedure done are very low with the biggest possibilities being bleeding and swelling which can be treated with medication.

Cosmetic Work to Make a Beautiful Bride Shine

If your wedding is coming up soon, now is the time to ensure that your smile is perfect along with your dress. The bride is the main event at the wedding, so all eyes will be on you. Since it is such a happy occasion, you will be smiling often and everyone will notice your smile – including the camera. By taking care of your mouth now, you won't have to worry or be self-conscious about your smile at your wedding.

The first step is to make sure that you take care of any oral health issues such as filling cavities and treating gum disease. You should have healthy looking gums and no dental pain by the time the big day arrives. You shouldn't have any more worries than necessary! Once these health issues are addressed, the fun part starts. Consider having your teeth professionally whitened so your dress won't be the only white glow to stand out. If time is limited, your dentist can do one long session for you to have radiant teeth in time for your wedding.

If your teeth could use some straightening and you still have six months to go, try our *Six Month Solution brace system*. With this

treatment, even if your wedding is quickly approaching, you can have straight teeth in time to walk down the aisle. Fill in any gaps or broken teeth with crowns or even implants and then make them look more natural by having your gums sculpted. If you still aren't happy with the fronts of your teeth, consider having veneers placed on. Finally, try to schedule your gum therapy as close to your wedding date as possible for an extra clean feeling and bright look. After you have taken care of your smile, send your groom in so he can receive a matching smile. Your wedding is the one day that you should indulge and look exactly the way you have dreamed of; so make sure your smile makes your dreams come true.

CHAPTER 4

How to Find the Right Dentist

CHAPTER 4

HOW TO FIND
THE RIGHT DENTIST

Though many dentists offer the same basic dental services, not everyone offers the same level of care, experience and cosmetic services. This is why it is important to thoroughly research a dental practice before you become a patient. After all, you don't want to have

a procedure done just to find out that they didn't provide the kind of service that you had expected.

First of all, you should understand exactly what a dentist does and what his responsibilities are. You probably already know that he provides oral care such as preventative services and treating oral problems. A dentist will check your gums, teeth and mouth to ensure that they are healthy. He or she are responsible for diagnosing all diseases and conditions of the oral cavity and will treat them accordingly. This is an important aspect of being a dentist because certain oral conditions could be indicative of osteoporosis, diabetes or even cancer, and early detection is of paramount importance here.

Your dentist will also educate you on how to prevent these conditions from developing. Typically, the dentist will not clean your teeth as this is the job of a dental hygienist. If a dentist practices oral and maxillofacial surgery, implants or administers general anaesthesia, he will need to have additional qualifications or training. If you are primarily interested in cosmetic services, you should ask your dentist if his clinic offers these services. At *Lifestyle Dental Clinic,* not only do we offer top medical care but also a plethora of cosmetic treatments that together can give you a healthy and beautiful smile.

Dentists are responsible for following a code of ethics and conduct so that patients can trust them to provide accurate and top care. Similar to medical doctors, dentists must keep with continuing educational requirements so that they are knowledgeable in changing dental recommendations. They are also responsible for keeping a clean and sterile office which is something that you should pay close attention to. There are many laws and regulations that dental clinics must comply with when it comes to cleanliness and safety standards. If they don't follow these regulations, then it could compromise the safety and health of their patients. At *Lifestyle Dental Clinic* we have our own sterilization unit which ensures that all policies are being complied with. We take the extra step to make sure that all of our equipment is clean and sterile. When you receive dental care here, you never have to worry about cleanliness because we make it a top priority.

Something else that you should look into is whether a dental office has payment options that work for you. Even if you have dental insurance, there are many dental procedures that you will have to pay a portion of the amount for. If you don't have dental insurance or you want to have cosmetic surgery, then you will have to foot the whole bill yourself. Since many people don't have the money upfront or they don't want to have to pay such a huge amount up front, you will need a clinic that offers affordable payment plans. However, some dental offices require upfront payment in whole or monthly payments that you cannot afford.

This leaves you unable to get the work that you need and want done, or leaves you stressed about affording the high monthly payments.

At *Lifestyle Dental Clinic,* we have a great financing option that really works with what you can afford. We typically can spread the cost of treatment across a period of 12 months or sometimes even longer with no interest charged. In addition, we will work with you to develop a monthly plan that you can actually afford. We would never want to see you go without the treatment you need or the procedures you desire simply because you can't afford the payment upfront. Once you have your financial worries put to rest, you can begin to focus on treatment and improving your smile which should always be your main focus in the first place.

Once you take in all of these considerations, you should next find a few dental clinics and make all your inquiries. It can be a great help if you have friends or family that can recommend their dentist to you or tell you who to steer clear of. Sometimes a dental clinic can look good on paper but when you talk to someone who has had treatment there, you can find out that their services are lacking. Perhaps they don't take the proper safety and cleanliness precautions as they say or appear to; or maybe the dentist is lacking bedside manner. No matter what, it is always beneficial to have first-hand feedback to factor into your decision.

If you can't find a dentist that you feel comfortable with simply by personal referral, then you should use an internet search engine to help you. First, start by searching for dental clinics that are convenient to your location. Many offices have their own website where you can see what type of services they offer, what their typical fees are, their finance options and plans, pictures of their office and sometimes even information and pictures of their staff. After you find a dental clinic that looks like it could be the right match for you, search for any reviews from their patients. You can often find reviews and ratings for dentists from people who have been treated there at least once. This can help you to form a complete opinion before you settle on your new clinic. If you have any more questions, don't hesitate to call the office and ask. Most dental staff will be happy to answer all your questions; and if they aren't, then you may choose to cross them off your list as an option. When you finally make an appointment you can further address any questions or concerns with the dentist. At this point, you should be able to decide if the dental clinic is the right fit for you or not.

Lifestyle Dental Clinic has its own website precisely for this purpose. We strive to answer all of your questions and any questions that you may have through our informative site. We even have photographs and bios for every member of the staff in our clinic. This way, you know exactly who you will be dealing with and their credentials so you won't have to be surprised. In the end, a dentist and his staff should keep your care in the highest regard and aim to keep you happy with their services, and if you can find a dental clinic like that then you have found the right dentist for you.

Parting Words

PARTING WORDS

Now that you have read my dental guide, you should be thoroughly educated on how to care for your mouth, what type of treatments are necessary, the cosmetic procedures available, the best treatments for your lifestyle and how to find the right dentist for your needs. Let's review the top points to remember:

1 Take care of your mouth now to save time, pain and money later.

If you take care of your mouth now, you can avoid advanced tooth decay or gum disease that will not only cause you a great deal of pain, but will also cost you money and time to have them treated. Avoid many problems by properly brushing your teeth twice a day, flossing every day, and seeing your dentist two times a year for an oral health assessment and thorough teeth cleaning. No matter what stage of life you are in or what your circumstances are, there are options convenient to you to take care of your mouth.

2 See your dentist as soon as you know there is a problem.

As soon as you notice any mouth pain, changes in the appearance of your teeth or gums, foul breath or any other dental change that alarms you, make an appointment to see your dentist immediately. Even with proper dental care, there are other factors that can cause dental problems including what you eat or drink, if you smoke, if you have any illnesses or diseases, if you take certain medications or if you have suffered any mouth trauma. Therefore, as soon as you notice a change, even if it is minor, see your dentist. It is always better to nip a problem in the bud before it grows into an even bigger problem.

3 After your mouth is healthy, work on making your smile shine.

After you have addressed any and all dental problems, you should decide what cosmetic treatments can improve the appearance of your smile. Your smile is powerful and can give you more inner joy than you may realise. If you are self-conscious of your smile, you just won't smile often enough. However, if you are proud of your smile then you will let your pearly whites shine, thus improving your mood and the mood of others around you.

4 Find a dentist that is right for you.

Having the right dentist is an important part of the process. He or she should be knowledgeable and practice the highest standards of safety, cleanliness and care. You should also be comfortable with the entire dental staff and trust your dentist's opinion and advice. Having a good relationship with your dentist will ensure that you get the treatment you want and need.

Now that you have all this valuable information about caring for your oral health and appearance, I hope that you will put it all into action. Myself and the entire staff at *Lifestyle Dental Clinic* have the experience and desire to help you achieve a healthy smile that you can be proud of. For something that is so important, don't put it off any longer. Call today so that you can get the smile that you have always dreamed of!

Dr Nadim Majid

Contact Information

Lifestyle Dental Clinic

Unit E, 284 Garstang Road

Fulwood, Preston, PR2 9RX

+44 (0) 1772 717316

info@lifestyledental.co.uk

www.lifestyledental.co.uk

 www.youtube.com/lifestyledental

 www.facebook.com/lifestyledental

 www.twitter.com/lifestyledental

ooking for more information about dentistry or a specific
cedure, you have come to the right place. My name is Dr.
jid and I am the Dental Surgeon and CEO at Lifestyle
ic located in Preston Fulwood. As a dentist I believe that
mportant aspect of the profession is communication. The
ommunicate the dental treatment to the patient in a
ey understand and then offer them some choices.

world of social media there are now many forms of
ation available. So, I started a blog, www.lifestyledental.co.uk/blog
series of videos through which I began to communicate
er some questions. These videos then led to this book.

ve shown that oral health impacts your overall health
is is a subject that everyone should be informed about.
preventative care and treatment is the most important
ntistry, you don't have to visit your dentist only when
oroblem or when you need a cleaning; in fact, there are a
f cosmetic procedures that you can undergo if you feel
e needs that extra step.

long with Lifestyle Dental Clinic, is to improve the lives of
nts by helping them maintain a healthy smile while also
their confidence. So please enjoy my book and expect to
mber of things that you never knew about achieving and
ng a healthy and beautiful smile.

Lifestyle
Dental Clinic

ng Road
R2 9RX
16
al.co.uk

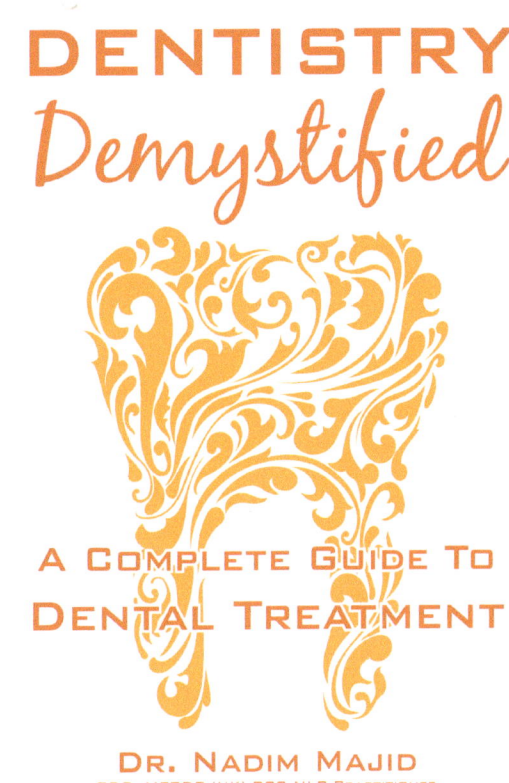

DENTISTRY
Demystified

A COMPLETE GUIDE TO
DENTAL TREATMENT

DR. NADIM MAJID
BDS, MFGDP (UK) RCS NLP PRACTITIONER

www.ingramcontent.com/pod-product-compliance
Lightning Source LLC
Chambersburg PA
CBHW041132200526
45172CB00018B/141